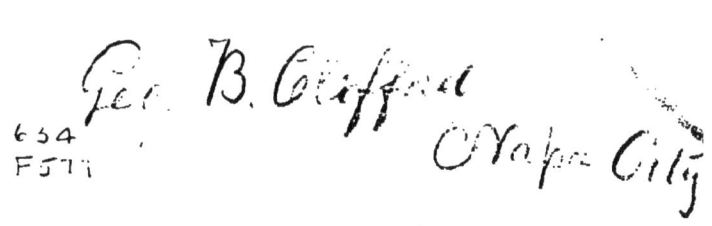

PRICE, ONE DOLLAR

A PRACTICAL TREATISE

ON

OLIVE CULTURE,

Oil Making and Olive Pickling,

BY

ADOLPHE FLAMANT,

OF NAPA, CALIFORNIA.

COPYRIGHTED 1887.

Louis Grégoire & Co., Booksellers, 6 Post Street,
San Francisco.

To REV. RICHARD WYLIE,
  Pastor of the Presbyterian Church,
    Napa, Cal.

DEAR SIR:—You take such a deep and intelligent interest in olive culture and I owe so much to your valuable assistance in the translation of this little work from the French—in which language I wrote it originally—that this, coupled with the high feelings of respect I entertain for your person, prompts me to respectfully dedicate to you this humble effort of mine in which it is my purpose to present to our agriculturists, from a Californian point of view, and in as brief and concise form as is in my power to make it, the principal points bearing on olive culture, which I hope and trust will soon stand foremost among the great industries of our highly favored State.

  Yours Respectfully,
    ADOLPHE FLAMANT.
NAPA, June 15, 1887.

THE OLIVE TREE.

CHAPTER I.

SOILS SUITABLE TO ITS CULTURE.

In the Book of the Hebrews we read that when Noah, desiring to ascertain whether or not the waters of the deluge were beginning to withdraw from the surface of the earth, dispatched a dove from the ark, that feathered courier brought back to him "an olive leaf plucked off." Ever since that time the verdant branch of this evergreen tree has been regarded as the appropriate emblem of peace and of abundance.

We read in The Book of *Judges*, chap. ix.: 8— "The trees went forth on a time to anoint a King over them: and they said unto the olive tree, Reign thou over us."

We also find in the Book of *Nehemiah*, chap. viii.: "And Ezra the priest said unto all the people: This day is holy unto the Lord, your God; Go forth unto the mount and fetch olive branches."

Amid the steep and arid slopes overlooking the valley of Jehosaphat rises the Holy Mount, clad in the constant verdure of the olive trees among which our Blessed Savior used to roam, and whence he wept over Jerusalem.

A venerable legend tells us that when Minerva became the tutelary goddess of the City of Cecrops, she caused the olive tree to grow spontaneously on the Rock of the Acropolis, which Neptune had just struck with his trident; and thence, kept with religious care, the olive tree became one of the emblems of the goddess.

Another legend represents Hercules introducing into Greece the olive tree which he brought from the Mounts Girapetra, in Creta.

Pindar says that all the slopes of Mount Olympus were soon covered with it, and that the Athenians used to crown with its branches the victors in the Olympic games.

From Greece it was introduced into Italy. There, says Virgil (Book II. Georgics), "the sterile lands and stony hills delight to be covered with the hardy and perennial olive tree."

Martial also says: "Hereafter the proud oak may be jealous of the olive tree of the Mount Alban!"

When the Phoceans founded Marseilles, about 600 years before the Christian Era, they introduced the olive tree in Provence, where they planted it only on steep and stony places.

In passing over a long period, during which the praises of the olive tree have been sung in many tongues, we come to that great agriculturist,

Parmentier, who during the last century was instrumental in the introduction of the potato into France. He says: "Of all the trees that the industry of men has made profitable, the olive tree deserves, without contradiction, the very first place."

It is thus on the Mount Ararat, on the mountains mentioned by Nehemiah, on the summits of the Acropolis, on Mount Olympus, on the Girapetra Mountains, on the steep declivities of the Holy Mount, on Mount Alban, on the rough and stony spots selected by the Phoceans, on the sterile and rocky places recommended by Virgil that the olive tree has found the kind of soil most propitious to its robust constitution.

Has the tree or the nature of the soil changed since so that we may wisely disregard the testimony of the ancients? Is there anything to encourage us now to plant the olive tree on a richer or a different soil from that of its historic and traditional growth?

In the works of the most reputed modern writers we find full and satisfactory assurance that the kind of soil cited by the ancients is the most favorable to the growth and profitable culture of the olive tree. I will here give a few citations:

Riondet, p 2. The olive tree requires a warm but temperate climate. It dreads equally extreme cold and excessive heat. In the north of Europe its progress is checked by cold weather;

in the south and in Africa it is stopped by too much heat.

It does not grow well on low and wet lands, but it succeeds perfectly on mountains as well as on hills, and even amidst rocks, provided there is but little soil; indeed, on places where there seems to be scarcely anything but stones.

What gives it still a greater value is that where it grows best other annual products are impracticable.

p. 13.: The olive tree is capable of enhancing the value of a soil naturally poor to a figure ranging from twelve to fifteen thousand francs per hectare. (Equal here to about $1,000 to $1,200 per acre.)

p. 42.: In France the finest and most productive olive trees are to be found in the neighborhood of the city of Grasse, and they are mostly always planted on steep and rugged hills.

p. 133.: The olive tree grows to perfection on dry lands, and in climates where often not a drop of rain falls for six and eight months.

Reynaud. p. 34. How much waste land could be utilized in the cultivation of the olive tree, which is so little exacting that it seems to be contented with a few baskets full of earth! In fact, where is the tree that, like it, would grow on arid, rocky spots, and without water? And then, it requires so little care, such slight cultivation, so little fertilizing!

p. 67.: The olive tree prefers hills and elevated places, does not grow well in valley lands, low and wet, or where water stands. Its fruit corresponds to the soil where it is planted; in rich and moist land it gives a heavy and fatty oil; in the warm and dry soils the oil will be finer; in the marly and clayey soils it produces less.

p. 106.: It can be easily understood why it cannot be planted in regular lines or rocky situations, where one utilizes even the interstices of rock, and where there is hardly enough earth to cover the roots.

Coutance. p. 177.: The best soils, where the fertility of cereals is so great, are not the proper ones for the olive tree, not that it will not grow in them, but because an exaggerated ligneous development will take place at the expense of the richness of its fruits. The quality of the oil will likewise be affected. M. Cappy establishes on this point the difference noticeable between the products of the olive trees of the fertile plains of Calabria and those of the stony hills of Lucca; between the products of the plains of Salon, in France, and those of the rocky hills of Marseilles and Montpellier.

The olive is the only tree for the arid, steep and rocky hills of the Mediterranean shores. The ancients knew this well. Mangon insists on the necessity of planting it on a dry soil. Lucilius Junior, Virgil, Columelle, and many others make the same point.

p. 139.: It will grow up to an altitude of 1200, 1800, and even of 2300 feet, in some instances.

Fabien speaks of the necessity of temperate climates for the olive tree. It prefers the neighborhood of the sea, and it is more fertile in proportion as it approaches salt water, to enjoy the sea breezes.

There are still at Jerusalem olive trees, that M. Bove estimates to be possibly over 2,000 years old, that have witnessed the great scenes of the Savior's Passion.

M. Enault affirms that he has seen the finest trees in existence at Mount Carmel, in Galilea and in Samaria, and that he has seen none anywhere presenting an appearance of extreme age so striking as those of Gethsemane found in arid and rocky situations.

Finally Delille affirms that he picked a branch of the famous olive tree of Athens, the age of which is admitted to be about forty centuries.

CHAPTER II.

REPRODUCTION.

The olive tree is reproduced in different ways: by the seed, by the simple cutting, by the ramified cutting, by suckers that shoot from the trunk, and by the woody excrescences which form on the bark of the upper roots of old trees.

Let us begin with the reproduction by the seed.

It must be first understood that an olive tree so grown has to be grafted, as it would otherwise remain a wild tree, giving thus but a poor and small product. On the other hand it is well known that through the medium of a seed a tree is more vigorous, has a more lasting power, resists better cold weather, and is less delicate on the choice of soil than those grown from cuttings. For all such reasons this is the mode most generally in use in the olive regions of Europe.

But when the olive tree is so robust by nature, so little scrupulous with regard to the choice of soil, enjoys such remarkable longevity, and has no excessive cold weather to fear in California, should it be raised by us from the seed instead of the cutting, when by the first mode we have to wait ten or twelve years for the product, as against four or five years by the second?

Moreover, grafting which becomes indispensable when the tree is raised from the seed, giving it thus additional vigor, can just as well, if so desired, be applied to the tree grown from the cutting without losing thereby the advantages derived from this last mode of reproduction.

Coutance, who pronounces himself in favor of the seed, tells us that the plant has to remain at least seven years in nursery, and that after being grafted it requires three more years before it begins to bear fruit.

Reynaud tells us also that he has seen in France, in the county of Ardeche, as also at Cannes and in the Hyera Islands olive trees raised from seed; that they were ready to be grafted, but that this result had required seven years. He however adds that the reproduction of the tree by seed has been found so slow that it seems puerile to have recourse to it.

Amoureux affirms that this method is of an excessive slowness and of very little practical use.

Charles Etienne and Liebault concur in saying that it is time and money lost to employ this method.

In Mr. Elwood Cooper's treatise on Olive culture we also find that when the tree is raised from seed it has to remain seven years in the nursery, but that when grown from the cutting it bears as early in Europe as it does in California.

Riondet explains to us how the young olive tree, raised from seed, develops always a long tap-root, which constitutes its principal and often its only support; and that when transplanting it to permanent site, after a long stay in a nursery, the cutting of said tap-root, which then becomes indispensable, inflicts upon its system a serious injury from which it is likely to suffer for years.

It seems thus established that the olive tree grown from the seed—which is the method most generally followed in the regions of Europe where the severe winters experienced occasionally make it desirable to render the tree as hardy as possible—has to be kept about seven years in nursery, and that at its transplantation it will experience a severe check which will be the natural result of meddling with its tap-root, as also of cutting back its top.

Is it then at all surprising that a half generation should pass before the olive tree so produced reaches bearing? Many people who have not carefully studied olive culture seem to believe that this is an inevitable result. We shall see by further explanations that it is not.

Let us pass now to the consideration of the propagation of the olive by cuttings. We would state in common with Coutance, Amoureux, Riondet, DuBreuil, Reynaud, and many others, that a cutting coming from an olive tree that has been grafted, and of a good variety, needs no grafting. This operation is however necessary when the cutting from a grafted tree is derived from a point below the place where grafting was effected.

These cuttings can be made like those of a vine or any other cutting, only with this difference that the olive tree being an evergreen, one or more sets of leaves should be left on.

It is difficult safely to cut the large truncheons because, when taken from the tree or even when cut a little to freshen the butt-end at plantation, there is danger of crushing the bark, which has the effect of imperiling their starting and which, should they grow, may induce rot. To plant them directly in permanent sites is to run the risk of losing a great many, as has happened to several parties I could name. If, on the other hand, they are placed in nursery in preference to much smaller cuttings, their tap-root will be so developed, even only after a year of stay therein, that it will be necessary to cut it back when they are to be transplanted, which will reduce their ultimate chances of growth and will at least make them languid and sickly for a year or two. But, the smaller the cuttings are when placed in the nursery, the less will be the chances at transplantation within a year of disturbing their root system which will necessarily be less developed.

These smaller cuttings, from six to eight inches long, are generally raised in boxes under glass, where they take very readily; or in open ground in nursery when from eight to twelve inches long; but there their growth is very precarious. When ready for transplantation within a year the whole root system can be taken with the soil adhering to it and placed in the ground without disturbing it, and especially without exposing it to the air.

I consider this last point of great importance, for it is well known that all evergreen trees, whose vegetation is nearly always active, are of a very difficult transplantation. The slightest exposure of their roots to the air renders the starting in their new places very doubtful. Any one who has had occasion to transplant eucalyptuses, laurels, orange trees, etc., must be acquainted with this fact.

In support of this theory I extract the following, from a recent article of the Phœnix *Herald*, giving a few sensible hints on the setting out of an orange orchard:

"The greatest care must be exercised in transplanting the orange not to allow the small thread-like roots of the tree to become dry, for the moment they do so the tree is gone. The roots must be carefully dampened till the tree is safe in the ground. This is one of the most important items to be observed in transplanting."

The olive is just as delicate to handle as the orange tree, so that the older it is and the more developed its root system, the more danger it presents in transplantation, when even the most careful precautions will not always secure success.

The small trees, when one year old, will develop with astonishing vigor when planted in their permanent sites; their tap-roots will sink rapidly; they will stand, without suffering, drought and hot weather, and not more than one in every two or three hundred will fail to grow. Not only had I occasion to verify this, but I have also observed that when so planted, without experiencing any amputation of their roots and branches, they will overtake in life and vigor before two or three years those which, planted older and larger, have had to undergo the mutilations which are rendered necessary by their greater age and a consequently more developed root system.

Mr. W. G. Klee, in a bulletin of the University of California, says that the mode of reproduction by large cuttings is liable to several objections. He claims justly that there will never be so fine a root system developed as by starting the trees from small herbaceous cuttings. He recommends to take from young, growing trees, the young tops, when neither very soft nor perfectly hard, having three to four sets of leaves, and to put them in a little frame with sand, where they are to be given a few waterings during the course of a month. He states that in three or four months the little cuttings will have rooted, and in a few months more will be found ready to set out. He adds that olive trees planted in the Santa Cruz mountains were propagated in this manner, that they received no irrigation after setting out, and that they have formed a beautiful root system.

Mr. Frank A. Kimbal, of National City, San Diego County, tells us also, that he has in no case succeeded with large cuttings, and that he has obtained but meagre results in planting twenty inches deep. He tried with all kinds of cuttings, from three feet down to eight and ten inches only, and he finds the latter preferable.

The mode thus recommended by Mr. Klee, by Mr. Kimbal and others, is in perfect harmony with what I have done, and which has enabled me to obtain an excellent root system in less than a year. Having had frequent occasions to compare it with others, I do not hesitate to pronounce it as the one method capable of producing most vigorous trees, which, within four or five years, will be from ten to twelve feet high, and will begin to produce a few gallons of olives.

I have knowledge of the fact that several persons have planted olive cuttings in nursery, and have met but with very meagre results. I think I can give them possibly the reason for it.

The olive, as already said, is an evergreen tree. It has two very distinct yearly vegetations, one called the spring vegetation, the other the fall vegetation. It is thus, that under our fine and quite exceptional climate, where the winters are frequently very mild, its vegetation knows scarcely any cessation. If the cuttings are not taken from the tree during one of those short periods of comparative repose—which vary according to seasons—and are not placed in the nursery within a reasonable time, say from one to two weeks, there is danger of the vitality in most of them dying out, and the loss will easily reach thirty, forty, or even fifty per cent., and possibly still more. In this respect, the cuttings of the olive tree differ from those of the vine, which can be cut immediately after the fall of the leaves, when vegetation comes to a stand-still, and which can be kept buried in the ground until March or April, without interfering with their starting when spring comes.

For the reasons here suggested, it can be understood why those who have attempted to reproduce the olive tree from cuttings which were not recently cut from the tree, and who have performed that operation at a season of the year when the sap was too active, have realized such poor results. I know of some parties whose loss has reached 80 and 90 per cent., and two of them who did not succeed with a single cutting. I can see no other cause for it than the one I have just mentioned. Let us now pass in review other modes of propagation.

Cuttings can be made from the suckers that grow from the base of the tree, but if they are taken below the grafting point of trees raised from the seed they will have to be grafted.

The olive tree is also reproduced from the woody excrescences that form generally on the trunk of old trees. This mode of propagation which carries with it the mutilation of the trunk of a tree is possible only in the countries where old trees are to be found, while from young trees, of which there will soon be plenty in California, cuttings can be easily procured through the ordinary process of pruning, which thus proves beneficial to them instead of being a source of mutilation. This alone should be a sufficient reason for the general adoption in this country of so rational a mode of propagation.

As much as possible a dry soil should be selected for an olive nursery. Riondet tells us that in irrigable lands a finer growth may be obtained, but when those young trees are transplanted to a dry soil, they suffer much, and it takes them several years before they get settled in their new place. An olive tree, weaker, raised in dry land, will always develop with more vigor than another one stronger coming from an irrigated soil.

The young plants in nursery, says Du Breuil, should be protected from drought only through the means of hoeings practiced during the summer, and he adds that when taking them from the nurseries they will accommodate themselves very much better to the burning soils where they will be planted than if they had been subjected to irrigation during their tender youth.

However, this practice which is recommended for the south of France and Italy where spring showers and summer storms are quite frequent and are generally sufficient to bring to those young plants the necessary elements of life, should not be adopted in an absolute manner in California, where sometimes there is no rain, or none of much account, from April until October or November. This is why, in the absence of rains, a few waterings distanced according to the season will perhaps be necessary to insure their start and promote their best development.

The little herbaceous cuttings raised in boxes, where there is not more than from four to five inches in depth of soil, should be occasionally watered, and just enough not to allow this thin bed to dry out, but it would prove a mistake to water them too abundantly.

For it might be said that the olive tree dreads too much water, or as much at least as will prove beneficial to other plants. While the vine cutting will grow luxuriantly with the help of repeated waterings, the olive cutting will suffer when similarly treated, and will certainly die out if its languid appearance, previous to its start being taken as indicating a need of water, it is too much soaked with it. Thus caution should be exercised in the waterings to be given to the young plants raised in nursery, but the ground should be loosened at their base as often as possible.

Frequent hoeing, while destroying the weeds, maintain always around the plants a moisture which is propitious to their growth, otherwise the ground would dry and form a kind of crust during the burning heat of the summer, especially in California, where rains are almost unknown from June to October; it would hardly be penetrated by the air and would not receive the beneficial effects of the atmospheric influences.

The most favorable season for transplanting to permanent sites those small one-year old rooted cuttings is dependent on the location selected. If the soil is light and dry, it should be done before winter; if heavy and damp, in the spring.

It is a fact generally admitted that in dry soils it is important to make plantations of rooted cuttings in autumn, their start being then much more assured; while in the wet lands, where the roots could not spread easily, it is preferable to wait until spring; for, the facility with which those soils retain the water maintains a permanent state of humidity which produces rot in the roots and brings about symptons of decay soon followed by the death of the plant. When on the contrary the planting is done in the fall in a dry and light soil, the water which it receives during the winter rains percolates freely to the lower stratum, the numerous ramifications of the roots, which under our exceptional climate know nearly no repose, absorb it through all their pores and the young trees develop with an astonishing vigor, gaining during a mild winter, as is so often the case here, from two to three months on the vegetation of the principal olive bearing zones of Europe, where winters are generally longer and more severe than those experienced in this perpetual-spring climate of California.

To resume what has been already said in this chapter, I am decidedly in favor of the propagation of the olive tree in our climate by means of small cuttings, freshly cut at those periods of the year when the tree experiences a comparative repose, coming directly or originally from grafted trees, and raised in boxes or nurseries. When one season old these little rooted plants are ready to be transferred to permanent sites; there they will make a growth of from two to three feet a year; they will develop more rapidly than vine cuttings of similar age, as I have had frequent occasion to verify, and they will begin to bear some fruit in their fourth year, as has been asserted repeatedly by Mr. Elwood Cooper of Santa Barbara; Mr. Frank A. Kimbal, of San Diego; Mr. W. G. Klee, of the State University; Mr. W. A. Hayne, Jr., of Santa Barbara; Mr. L. A. Gould, of Auburn; Mr. Isaac Lea, of Florin; Captain Guy E. Grosse, of Santa Rosa; Mr. H. W. Crabb, of Oakville; Mr. A. B. Ware and Colonel Geo. F. Hooper, of Sonoma, etc.

In reference to this I will quote the following extracts:

From Mr. Elwood Cooper's Treatise on the olive: " Trees growing from cuttings will produce fruit the fourth year, and sometimes, under the most favorable circumstances, will give a few berries the third year. My oldest orchard was planted

February 21st, 1872. At four years I gathered from some of the trees over two gallons of berries. In 1878 over thirty gallons each off a few of the best trees, the orchard then being only six years old.

"The newness and richness of our soil will probably give, the first fifty years, double the best results given in the oil countries of Europe."

From the San Francisco *Evening Bulletin*, May 3, 1887: "It is often stated that the olive tree will not bear until its seventh or eighth year. Capt. Guy E. Grosse, of Santa Rosa, thinks that on his mountain ranch may be found proof satisfactory that such is not the case. Out of his 500 olive trees, planted four years ago, 461 are in full bloom and promise a good yield. Further and successful refutation of the statement concerning the maturity age of the olive may be found in this city. On A. B. Ware's premises is an olive tree four and one half years old which bore a good crop last year.—*Sonoma Democrat*."

Similar affirmations have been made by so many other parties who have engaged in olive culture in California that it seems unreasonable to doubt it any more. It leads us to the belief that those who have olive trees here bearing nothing, or but a few small berries, at an advanced age, must have reproduced them from cuttings taken from trees raised from the seed and which were never grafted.

That the fact of the early bearing mentioned

above is the result solely of the mode of propagation by cuttings, of our exceptional climate, or of our virgin soil, or the result of these three elements combined, matters but little. It is a fact, and the proofs are superabundant.

But if, on the contrary, the olive tree is raised from the seed it has to remain many years in the nursery until it develops sufficient strength to admit of successful grafting; and when transplanting it, the tap-root as well as part of the top having to be taken off, the roots also suffering from exposure to the air which it will be impossible then to avoid, all of this will combine to inflict a long and severe injury upon its whole system, which will delay considerably its bearing period.

There will certainly be a difference of fully five years in the time of its production according to the one of those two modes of reproduction that will have been adopted.

CHAPTER III.

GENERAL CARE.

If, in the cultivation of the olive tree, one were to be guided by the ancient beliefs that have come down to us through the ages, it would appear that when once planted it can be left to take care of itself.

Virgil says in his Georgics that the olive tree needs no cultivation, and Pliny repeats with him that it should not be given too much care.

Columelle affirms also that of all trees the olive is the one which requires the least work and the least manuring. He does not, however, recommend an absolute abandonment of "the first of all trees," as he calls it, but judges that it is the tree *par excellence* that can stand neglect and bad treatment better than any other.

It has nevertheless been since recognized that the olive tree, though by no means exacting, needs a certain amount of care, especially as regards pruning. It might be said, however, in connection with this, that in certain olive regions of Europe, Africa and Asia, there are still many olive trees that are never pruned and receive no care whatever.

It is thus that we read in Dr. Thomson's "The Land and The Book:"

"This tree requires but little labor or care of any kind, and, if long neglected, will revive again when the ground is dug or ploughed, and begin afresh to yield as before. Vineyards forsaken die out almost immediately; and mulberry orchards neglected run rapidly to ruin; but not so the olive. I saw the desolate hills of Jebel-El-Alah, above Antioch, covered with these groves, although no one had paid attention to them for half a century. Large trees, in a good season, will yield from ten to fifteen gallons of oil. No wonder it is so highly prized."

Reynaud tells us that in the south of France the olive tree gives abundant product without the effort of a careful and costly cultivation. "Which is the tree," says he, "which, like it, demands so little care, so little cultivating, so little manuring!"

Other modern writers, on the contrary, insist that the olive tree should be carefully worked, pruned and manured.

Between these two extreme views it is well to allow ourselves to be guided, to a certain extent, by the experience of past generations, which is often transmitted to us under the form of proverbs. It is thus that we have been cradled in our younger days by such old sayings as: "An olive tree requires a wise man at its foot and a foot at its head;" and yet: "Make me poor and I will make thee rich;" from which we should see that the tree is not so much in need of a costly stirring of the soil as it is of a careful pruning.

The caution that is thus recommended to us, as regards to the cultivation of the soil around the olive tree, is in a certain measure the natural consequence of the rocky and steep situations where it is most generally found in Europe, and where the plows cannot find easy access. In such places, where plowing is out of the question, two or three hoeings a year, a few feet around the tree, will be found sufficient to ensure its rapid development. Eugenio Ricci says to this effect: "The soil should be dry and stony, and on a slope. There should be no other cultivation except occasionally to remove the grass and loosen the soil. At least twice a year the land should be worked with the hoe for three feet around the tree, which process should, every second year, be preceded by a manuring."

If olive trees are planted in arable lands, then the heavier the soil the oftener it has to be stirred, while on light soils it can be done less frequently. It is thus evident that the cultivation of the olive tree should not be identical in all soils, and it belongs to each olive grower to apply the most suitable method as per the character and constitution of his land.

Manuring the olive tree meets with no opponent, for no one could ignore the advantages it presents.

As for pruning there are many divergent opinions. An olive tree never pruned bears heavily one year, and gives but little fruit in the year following, as if it needs rest for its laborious efforts; but by judicious pruning it is brought to give regular yearly crops.

Du Breuil tells us on that subject that the berries of the olive that is not pruned are very numerous, and that they remain on the tree until the end of winter, so that during the fertile years all the sap has gone to supply their growth preventing new bearing branches from forming for the following year. It is thus that the fructification of the olive trees not pruned is most always biennial.

The pruning of the olive tree should have mostly for its object to decrease the height of its head so as to render the picking of the crop more easy; to give to that head such a form as to allow light and ventilation in all its parts; to suppress every year a certain number of the bearing branches so that the sap can feed better those that remain, and that by the development of new branches it may assure a good average crop every year.

Young olive trees are generally left to themselves for the first two years following their planting, pruning being applied only in their third.

Riondet recommends to direct the tree in such a manner as to avoid the ultimate necessity of having to suppress a large branch or to inflict a big wound upon it. It will be sufficient to that effect to clear the tree of the small branches that can no more bear fruit.

Contance guards us against the unreasonable pruning that seems to be recommended by the proverb " a wise man at its foot, a fool at its head," though he would rather prefer it to a complete abandonment. He simply recommends the suppression of all dead wood, the cutting of the branches that prevent light and air from circulating into the center of the tree, the giving it a regular shape, and the keeping it from growing too high, which would result in the sterility of the lower parts and would render the gathering of the fruit more difficult.

The suckers that grow continually from the base of the tree should be removed at intervals; and, while pruning, it should be borne in mind that the horizontal branches and those turning down are the most productive.

Considering the heavy summer winds experienced on the Pacific Coast it is highly advisable to form the trees low; they are thus less likely to be damaged and can be cleaned with washes against insect pests with more facility. Moreover, by keeping them so, the trunk develops with more force, the crops come quicker, are more abundant, and the pruning as well as the gathering of the crops is easier and more economical; all reasons that speak in the most eloquent language in favor of this protective and beneficial method.

CHAPTER IV.

COST OF A PLANTATION.

It should not be understood by the numerous quotations given in preceding chapters that the cultivation of the olive tree is impossible or will give but meagre results in a rich soil, well cultivated and abundantly manured. Such a soil, well selected, and provided it is properly drained, will give a good yield, which will be, as in all other cultures, the direct result of the good care that will be given to the tree. We should, however, bear in mind that it has been said time and time over by the best authorities on the subject, and especially by Michaux, that the quality of the fruit of the olive is essentially affected by that of the soil, and that while it succeeds in good loam capable of bearing wheat and vines, in fat lands it yields oil of an inferior flavor and becomes laden with a barren exuberance of leaves and branches.

Moreover, those rich valley lands are not always within the means of all parties who desire to avail themselves of the numerous advantages that the culture of the olive tree presents. Those with but modest means at their disposal will rather invest in fifty to one hundred acres of hill lands, more or less rocky, if their outlay for the same will not be above that required for the purchase of from ten to twenty acres only of a richer soil, especially if it is fully demonstrated to them that the olive tree will grow well on those steep and stony lands, and that while they are not likely to lose anything in quantity, relatively to a richer soil, they will gain the advantage of a finer quality in the product.

Let us thus study the cost of a plantation and its proper care on rocky lands, the price of which may vary from $10 to $30 per acre, according to their nearness to or remoteness from a city, or facilities of transportation. I will take in this as a basis, my own plantation of about 6,000 olive trees, which I made in 1884 on hill lands, most of them inaccessible to the plow, and where I have had all the work done by hand.

Planting as much as possible at a distance of twenty feet—for on such places regular lines could not be made—we have about one hundred trees per acre, leaving out the very few patches where it is utterly impossible to plant even an olive tree. We will select for this plantation, one year old rooted cuttings, coming directly, or originally, from trees that have been grafted, of which the stem will be hardly from ten to twelve inches high, and the roots from three to six inches long.

For such small trees it will be sufficient to dig

the ground one foot deep, or one foot and a half when the soil permits, but where the hole cannot be dug as deep as that, some of the surrounding earth can be brought around the tree in a concave shape.

The digging of these 100 holes, and the planting of the tree, should not cost above $5 per acre. Two hoeings of a space about three feet wide, around each tree, one in early spring, one in early summer, at $1.50 each, will make it $8 altogether per acre. The small rooted cuttings can be had at prices ranging from $10 to $15 per hundred, according to sizes; and taking their maximum cost of $15, we come to a total of $23 per acre for all the first year's expenses, independently of the cost of the land, which can be bought as cheap as $10 per acre, and even cheaper, if the purchaser is not particular about being near a city or a railroad.

During the following years, three hoeings, distanced according to a more or less rainy season, will be more than is required to keep the plantation in very good condition; it will not cost altogether over $5 per acre, to which can be added the cost of pruning every two years, and, if desired, the cost of manuring every two or three years.

When comparing this simple and cheap work with the care required by a vineyard, which, besides the regular cultivation, pruning, plowing, cross plowing, hoeing, tying, needs expensive stakes, suckering, summer pruning, sulphuring, etc., one can readily perceive the advantages to be found in olive culture.

CHAPTER V.

DISEASES.

The olive tree is so robust by nature, and its bark, leaves and fruit are so bitter that in consequence of those different advantages it is less exposed than other trees to the ravages of insects and animals, especially when it is planted on hills and mountains, on light and well drained soils, for, no one can ignore the fact that fruit trees in general are so much more exposed to the many pests that endanger their existence and check their bearing capacity as they are planted in low and moist lands.

The most dangerous enemy of the olive tree seems to be the black scale. This insect has a marked preference for the orange tree, as well as for the laurel tree, which are generally planted in rich soil and very seldom on high elevations or meagre and well drained lands.

Under the shape of very small shells of a dark brown color, these insects fasten themselves very closely to the branches, leaving after them a trail of a blackish dust formed by the sap they extract from the tree mixed to their leavings, the whole of which is covered by the dust spread over it by the wind.

Riondet tells us that during the winter, when the young insects are still under the calapash of the mother already dead, under which they remain yet protected from the cold weather, a large number of them can be crushed by rubbing the branches with a hard brush dipped into vinegar.

Lardier recommends to rub with lime water; Reynaud says to sprinkle the tree with that same preparation, and Du Breuil affirms that such an energetic rubbing and spraying will cause the disappearance of both the insects and their black trails.

In California where the black scale has pretty generally appeared and caused great havoc among the orange trees of the southern part of the State, the trees when young should be sprinkled with a whisk broom dipped in a bucket containing a mixture in equal parts of sulphur and whale oil dissolved in hot water; and if any olive grower is ready to act the moment those insects begin to appear here and there on any of his trees, he will easily prevent their spreading to others.

Other preparations in which enter kerosene, or infusions of tobacco, absinth leaves, etc., are also recommended, but while I fully appreciate their merit, the one just given brings about very satisfactory results; it can be prepared very easily and will not cost over ten cents a gallon, which quantity will be amply sufficient for the sprinkling of fully one hundred young trees.

But neglect and ignorance are capital sins in most things of this world, especially in arboriculture, and I have seen olive trees, in rich soil, adjoining orange, lemon and laurel trees, none of which had ever been cleaned and which were covered all over with black scale. How can any one expect a product from trees that are in the clutches of death? If, however, in spite of their leprous condition, they give a little fruit, it can be but in very small quantity, and if allowed to grow so, their weak and sickly condition will delay, if not preclude their propensity to bear.

We can thus say with reference to this that an ounce of prevention is worth ten pounds of cure.

There are other insects of a secondary importance which attack the olive tree, but provided one keeps his eyes opened to face the enemy when its vanguard puts in appearance there is but little to be feared.

Moreover, the ever provident nature comes oftentimes to the rescue of human carelessness. If the insect known under the scientific name of *Hylesinus oleae* bores the young twigs to the heart, causing some of the branches to break down under a heavy wind; and if a fly called the *dacus oleae* deposits its eggs on the berry and feeds afterwards on its flesh, the ant, and other insects belonging to the carnivorous class come and feed on them.

But the best preventive is, first, to prune carefully, so as to give free access to light and air in all parts of the tree; then to rub energetically the trunk and largest branches with lime-water, or with the whale oil and sulphur preparation, and these enemies will be kept at bay; only a small portion of them, if any, will survive, to become the prey of their own enemies, which nature provides to finish the work of man, that is: of the man who knows how to help himself.

In Europe, says Du Breuil, the most dangerous enemy of the olive tree is the excessive cold of certain winters, when the thermometer runs down as low as 12° and 10° Fahr.

Looking back over a century, it has been found that the olive trees have been frozen, on an average, every nine or ten years. Among the winters most disastrous to them were those of 1740, 1745, 1749, 1766, 1770, 1789, 1795, 1811, 1820, 1830, 1837, 1843, 1859 and 1866. No more recent data is at hand.

Is it, then, surprising that this culture rather tends to decrease than to increase in the best oil regions of Europe, and that they should raise the olive tree mostly from the seed? It is true that it can thus cope more successfully against those severe winters which come at frequent intervals; but the period of bearing is considerably delayed thereby. It is for such reasons as these that many European agriculturists are deterred from engaging in this culture, the result being that the production of *pure* olive oil is far from being equal to the consumption of the world; hence the many adulterations in which cotton-seed, linseed, cocoanut, lard oil, etc., play such an important part, and tend to keep up a natural prejudice against an article whose bad taste is so far from the exquisite delicacy of the genuine.

But similar dangers are not to be feared in California, where the thermometer falls very seldom below 30° Fahr., and will reach only very exceptionally 28°, or one or two degrees lower; while the Olive tree will stand without danger as much as 15°, and even 12° and 10°. It is thus that what is a serious check to its development in Europe, is the very reason why its culture should be adopted fearlessly and extensively under the temperate climate of California, for which Providence has been so lavish in its beneficent gifts.

CHAPTER VI.

VARIETIES.

We are told by Coutance that the primitive type of the *Oleaster*, or wild olive tree, has been modified in many manners, that numerous varieties have sprung up, that the nomenclatures prevailing in different localities do not correspond with each other, that it, therefore, is impossible to give a general catalogue which would comprise all the cultivated varieties of the olive tree.

Other authorities on the subject enumerate varieties in vast numbers. One writer will indicate certain ones not mentioned in another, and some of them, not satisfied with the varieties generally known, seem to take the task of discovering new ones, after the manner of an astronomer in quest of new planets. Moreover, the names vary according to the country, and it is often the case that different olive trees are designated under the same name. When thus the high priests in oleiculture have admitted the impossibility of giving a complete catalogue of the innumerable varieties of the olive tree, how could I dare to undertake so arduous a task?

I judge it then more practical to confine my attention solely to the varieties already most generally known in California, that have been acclimatized here after many years of cultivation, and I shall simply cite all that I have been able to learn of their respective merits, leaving it to more daring writers to recommend better ones among the great list of those known in all the olive regions of Europe, Asia and Africa. Let any one who will feel so inclined experiment with some of these latter ones, as regards their adaptability to our soil and climate, and wait years and years before realizing whether or not they will give better products in greater abundance and in shorter time than those that are already known to us.

Why should it be different with the olive tree from what it is with the vine? Who ignores the fact that in the wine districts of Burgundy, of Champagne, of Bordeaux, and in other places, vineyards in immediate proximity to one another, cultivated in the very same manner, and planted with cuttings belonging to the same variety give wines of a different character; while one will be considered of an ordinary quality the other will rank among the most renowned. Will the combined influences of soil, climate and exposition, which are of great importance for the products of the vine, work in a less degree for those of the olive tree?

Moreover, while planting the varieties which

already well known in California if, in years
to come, it is satisfactorily demonstrated to us
that better ones have been acclimatized, it will always be in order to use them for grafting our
trees after the experiments, which are generally
very costly in agriculture, will have been made
by those who have time and money to risk in
that beneficent manner.

We have therefore at present these three
varieties pretty generally known: The Picholine, the Mission and the Queen, or Reyna. We
will take them by turn and quote what the
writers most reputed on the subject have to say
of them.

Reynaud. Picholine, called also Colliasse.
This variety was named after Picholini, of Saint
Chamas, France, an intelligent agriculturist of
the last century, who was the first to graft the
Sauvageon on the Saourin and obtained such
good results therefrom that a sort of enthusiasm
seized the whole country in favor of that practice
which has been quite generally followed ever
since. In the Gard district, from the plains to
the top of the mountains, even in the fissures of
the rocks, every spot where there is but a little
vegetal earth is covered with this variety of the
olive. It should also be said that the Picholine,
amongst all other varieties is the one that seems
to be the least subject to the attacks of insects.
It is known to bear in much greater abundance
than the more common trees of the country.

Du Breuil. Picholine, called Saurin at Nimes, Sourenque at Aix, Plant d'Istres at Beziers: oil very good. The fruit is the best among those for pickling. The tree is very fertile.

Coutance. Picholine, alias Piquette, Saurins Coiasse, Plant d'Istres, Lechin: variety cultivated mostly in Provence, France. Oil fine and sweet; esteemed for pickling.

Michaux. The Picholine gives the most celebrated pickled olives. This variety is not difficult in the choice of soil and climate.

Pohndorff. Picholine, also called Lechin, Cuquillo, Olea ovalis, oblonga, Taurine, Plant d' Istres, Collias and Coias, known as the fine, sweet-pickling fruit bearing tree, which received its name from an agriculturist of last century of the name of Picholini. This tree is little damaged by insects. The fleshy olives, which stick to the kernel, are of red color when ripe, yielding a very good oil, and for pickling green, excellent. This tree resists in cold regions up to 14° C. below zero (about 7° above zero Fahrenheit).

Bleasdale. The best olive for pickling is the Picholine (olea ablonga). It is also valuable for oil.

W. G. Klee. The Picholine is a very hardy and rapidly growing variety.

Bernays. The Picholine, alias Colliasse, is known in France, Provence, as the best olive for pickling. It is among other choice varieties for oil. This tree is amongst the most productive kinds and possesses the additional advantage, in common with a few others, that it never grows large, thus the fruit is easily gathered.

Let us add to the credit of the Picholine that the much lamented Mr. B. B. Redding, while in Europe many years ago, studied most carefully the question of the olive tree. After many careful researches and comparisons he pronounced in favor of the Picholine as the variety that seemed to be most likely to give the best results in the California soil and climate. It is to him mostly that we are indebted for having this most excellent variety among us.

Let us see now what has been said of the Mission.

Pohndorff. The California Mission olive is the Cornicabra Cornezuelo variety, which requires more heat than any other. In the northern oil zone of Spain, the Cornicabra tree of great size is called Azebuche, or wild olive tree, for the reason that the fruit does not ripen there. The regions of Saragossa and Salamanca in Spain, are not warm enough to allow the fruit of the Cornicabra —known in California under the name of the Mission—to mature. In certain parts of our State, at San Diego for instance, the fruit of the Cornicabra ripens as early as the end of October.

W. G. Klee, of the University of California, tells us that when the mission fathers first landed in California, they brought with them two varieties of olives, one of which especially has been propagated throughout the State, but that although a most excellent and hardy variety, is not here as it is in Spain, adapted to the warmer parts of the country only.

Gustav Eisen, the well known vineyardist of Fresno, who has planted both the Picholine and the Mission, says: The Picholine seems to do well, is easily grown and transplanted, but the Mission I consider as less valuable. The first year when transplanted it generally loses all its leaves. It grows only very poorly from cuttings, and bears only when six to seven years old.

H. N. Bollander, who had charge of the botany of the geological survey of the State, and John Ellis, of the horticultural department of the University, have reported that the Mission olive is a shy bearer.

Major Utt says that the Mission olives will ripen two months later than other European olives.

As per the Queen olive, Reyna:

Bleasdale says that it is of very large size and is pickled for eating. The tree of this variety produces but little fruit, and the fruit when pressed yields very little oil.

Coutance. Spanish olive; large berry, oil bitter, esteemed for pickling.

After the aforementioned quotations is it necessary to give an additional reason in support of my belief that the Picholine ranks among the most desirable varieties for California? I was born in the oil regions of France, where the Picholine reigns supreme. I was saturated, I might say, from boyhood to manhood with Picholine oil and Picholine pickled olives. On my arrival at Napa, and while visiting its beautiful valley and the surrounding sections, I soon realized the correctness of the reports I had read about its climate compared to that of the south of France and of northern Italy, a very exact confirmation of which was given lately by Mr. Albert Sutliffe in the following words:

"The citizen of California who travels in Italy and the south of France cannot fail to remark the similarity of soil, climate, conformation of ground and general atmospheric conditions to those to which he has been accustomed on the Pacific Coast. In the vicinity of Marseilles the summer is almost absolutely rainless, while the winter rains are copious. The heat of midsummer is warm, but generally tempered by sea winds." It is thus that, guided as much by the sweet remembrances of the past as by the careful studies I made of the subject, I did not hesitate to adopt the Picholine for my own plantation.

Let us see now where the olive oils most reputed come from.

Elwood Cooper's Treatise on olive culture: Extending from the promontory of Saint Tropez, in France, to Lavonia, in Italy, in the gulf of Genoa, Nice is situated, whose reputation for the best oil has succeeded all other places in the world.

Bernays. The finest kinds of oils have hitherto come from Provence, in France, and Lucca, in Italy; the commoner from the Levant, Mogador, Spain, Portugal and Sicily. The olive tree will thrive and be most prolific in dry, calcareous, schistous, sandy or rocky situations; it will bear sooner and be more prolific than if grown in the rich soil.

Coutance. The oils of Spain are of a very inferior quality, especially when compared to those of France and Italy. They sell at a lower price than the latter.

DuBreuil. If the olive tree does not thrive well in a cold climate, it fares no better in very warm regions. It has reached great size at Cayenne, at San Domingo, but it never bore fruit there.

Bertile. The African coast produces a very inferior article, which can only be used for lamp oil or grease. Some of the Islands of the Grecian Archipelago and the western shore of the Adriatic produce better oil, but destitute of sweetness and suppleness, qualities most desired by consumers, and only found in the oil made in the valleys south of the Alps.

Dunham J. Crain, American Consul at Milan. The best article is produced in moderately warm regions. Thus, the oils of Italy are more esteemed than those of the Orient, and of the former, the oils of Pisa, Lucca, and San Remo are better than those of Sicily and the Neapolitan provinces.

G. Saint James: Pliny awards the palm to the

olive, the oil of which was at least more —— than that produced in the western coun—— So far as regards the oil of Spain, and to —— extent, that of Italy, this judgment holds —— to the present time, for the reason that the —— olive is a larger and coarser fruit, while —— Italian growers are too apt to detract from the —— delicacy of the virgin oil, by the sacrifice —— quality to quantity. It is almost impossible —— any one to realize the exquisite delicacy of the —— expression of the freshly gathered olive, un—— he has sojourned in such a district as that —— which Avignon, France, is the center.. The —— of Aramont, in Provence, was formerly supposed to have no equal in Europe. The oils produced in the south-east of France remain without a rival among those of the whole world.

The numerous quotations I have just given seem to demonstrate in a conclusive manner that the larger the fruit, the coarser is its quality; also, that the olive tree requires a temperate climate, and that an excess of heat acts unfavorably on the fineness of its product.

From this, we can infer that the oil that will be made in the Napa, Sonoma and Sacramento valleys, whose climate is very similar to the one of Provence, in France, and Tuscany, Italy, will be of a finer grade than that produced in southern California, where the climate presents more striking analogy to that of Spain, Portugal and Sicily.

It can also be inferred that the varieties of the

olive tree so reputed in Provence, and among which the Picholine holds a most honorable place, are those that will grow best on the hills of Napa and regions surrounding, while the Spanish varieties, to which the Mission and Queen belong, will be best adapted to the southern climate, where, however, they will give, as they do in Spain, a common product.

Bleasdale seems to confirm this theory, when he states that the olive tree flourishes in Egypt, Arabia and Persia, but that no one seems to dispute the poor quality of its product.

It is but just to state that a good oil is already produced in California, made out of the Mission variety; but it is made pure; it is so given in a virgin state to consumers, and this is the main cause of its merit, and of the great demand it enjoys. It thus gains by comparison with the trade oils that come from France and Italy which are mostly liberally mixed with cotton-seed oil, linseed oil, lard oil, etc.

We read in a Consular's report from Milan, that no unadulterated olive oil is exported from Italy, and statistics show that not enough genuine olive oil, fit for table use, is produced to supply the wants of the world. Much that is sold as olive oil, is the oil from cotton seed or sesame seed. Hog's lard is shipped to Italy from America, and comes back in bottles labeled "olive oil." These facts have an important bearing upon the question of future profits from olive groves in California.

Albert Sutliffe wrote lately on that subject from Florence: "Any one who has eaten the olive oil commonly used in America, and has also tasted it pure at the refineries of Nice, Lucca or Florence, can easily understand the prejudice against it. The two articles are entirely different. The former, too often suggests whale or lard oil in some state of impurity or rancidness, while even the most prejudiced person tasting the latter at the place of production finds it pleasant. Even the most fastidious or uninitiated taste would not object to a beefsteak cooked in the best of Lucca oil, which he would hardly be able to distinguish from the finest butter."

These adulterations are not confined to the shipping points of Europe, they also take place at many receiving centers where they dilute still more what has already been quite liberally adulterated. In reference to this, we just read in the San Francisco *Evening Bulletin* of May 29, 1887: "The 'Camera de Commercio Italiano,' an organization of local Italian merchants formed to promote trade between California and Italy, will hold a meeting Saturday at the rooms on Battery street, opposite the Postoffice. Various matters of business will come up; among other things, that of the recently discovered adulteration in this country of certain Italian products. It has been learned that some local dealers of more enterprise than integrity, have been importing olive oil and diluting it here with a cotton seed product.

The bottles were then recorked and resealed, and the expanded oil sold at trade prices as the genuine Italian product." At said meeting a resolution was adopted instructing the Secretary to notify Secretary Rubini that the chamber would do what it could in suppressing the evil, but that it was almost powerless to act for the reason that no California law prohibited the importation of oil, no matter what its character may be. This memorial, forwarded to the Italian Government, requested also that no adulterated goods be allowed to be shipped from Italian ports to this country.

I will then say in conclusion of this chapter: let us select such varieties of the olive tree as are universally acknowledged to rank among the best, both for pickling and oil-making; let us choose only temperate regions, so that our trees cannot be affected either by too much heat or too severe cold weather; let us plant them on the dry and sunny slopes of our hills where quicker production and finer quality will be obtained; and when, as a natural result, we make and serve to consumers a pure oil we can compete successfully with the foreign adulterations in all the markets of the world, if we can ever begin to supply our immense home consumption, which alone will absorb all that we can produce for many generations to come.

CHAPTER VII.

THE OLIVE OIL.

The quantity of oil in the olive berry, says Du Breuil, keeps increasing until the very last moment it is picked from the tree.

According to this theory, instead of gathering the olive in November or December, when it is already ripe, a larger quantity of oil can be extracted from it by waiting until February or March.

It should however be understood that the early gathering of the berry gives a better quality of oil, so that it is only where quantity is more desired than quality that the picking of the crop is delayed; but if the finest quality of oil is aimed at, then the gathering should take place either in November or December, so soon as the berries are ripe. After having picked those that have fallen naturally to the ground, the others are detached from the tree either by hand or by knocking the branches with long poles. They are then carried to the barn, where they can be crushed and pressed at once if desired, though most generally they are spread in thin couches and turned over with a wooden shovel, once a day for a week or so, in order to keep them from moulding; after which they can be crushed without further delay, or be placed in sacks in which they can either be kept for a while longer, or be shipped to an oil manufacturer if the grower has not secured the simple appliances required for making the oil himself.

I have made two plain sketches of a press and a crushing stone, which will be found annexed to this work. They are far from being as perfect as they might be, but I never was very proficient in the art of sketching, and mean by them only to give a general idea of the simplicity and cheapness of the machinery and appliances with which I have seen oil made in the old country. Steam power is used in most of the large oil-mills of Europe, but farmers with a few hundred or even a few thousand trees can understand by my sketches the kind of apparatus they need, and the facility with which they can take care of their own crops.

The first one of those two sketches shows the kind of rolling stone employed in crushing the berries. Set in motion by a horse, it revolves in a circular trough in masonry, or wood, in which the olive berries are thrown, discarding those that may have fermented; a thick paste is thus obtained, which is placed in straw bags of a circular shape, open at the top, which are then piled up from eight to ten at a time under a press in the style of the one represented by the second sketch.

The first pressing, made slowly and gently,

gives what is generally known as Virgin Oil. The bags are then removed from under the press, the paste is stirred up, boiling water is added, and a second pressure, harder than the first one, gives the oil that is most generally sold under the name of Virgin Oil, and which is still of a very good quality. The oil floating in the receptacle above the water is skimmed off with a large concave sheet of copper or tin.

The same paste, to which the fermented olives are added, with plenty of boiling water, is pressed once more, as hard as possible this time, and an oil of an inferior quality is obtained, which is used mostly in the manufacture of soap, of broadcloth, for lighting or lubricating purposes, etc.

This last operation is generally performed with a different press than that used in the two first pressures, so as to prevent this lower grade of oil from communicating a bad flavor to the better qualities. It is also considered highly essential in the extraction of the better grades to employ only apparatus of perfect cleanliness, and receptacles that are not used in the preparation of the oils of inferior quality.

When all the pressing is over, the paste left to dry is then cut in pieces and is used for fuel, for manuring, as also for food for horses, cows and other farm animals who are fond of it, and who fatten rapidly when fed with it.

The oil, placed in tin tanks, will deposit its impurities by natural rest within a month or so, when it can be drawn off into other cans or packages for the trade.

But, this mode of refining by natural process can be hastened by filtering in cylindrical tin vessels, with cotton batting at the bottom, in which case it can be bottled and sold immediately after. It has then, when just made, a freshness and delicacy of flavor which does not exist to an equal degree in the older product, which gains only a finer color by time.

Decandolle estimates the quantity of oil produced by the olive, at fifty per cent of its weight.

Sieuve says that one hundred pounds of olive berries will give about thirty-two pounds of oil, while other writers give an average proportion of product of twenty-five per cent.

It is, however, proper to state that this proportion varies, naturally, according to the variety of the olive. Some of an inferior quality are known to give as little as fifteen, and even ten per cent. On the other hand, it will vary according to the early or late picking of the crop, for, as I have already said: if you wish quality, pick early; if you wish quantity, pick late.

USES OF THE OIL.

The Scriptural books teach us how the olive oil was considered as a symbol of the divine grace, and, consequently, the important place it occupied in the religious ceremonies of the

Hebrews. A person anointed was considered as sacred. Oil signified unction itself, and he that had received it was consecrated king, priest or prophet.

The use of the oil in the Roman Catholic Church is too well known to need special comment. The Christian nations kept up the same traditions, which, from Saul to Charles the Tenth, of France, have hardly known any interruption. It is thus that we find the oil in the sacred lamps of churches, in the administration of Sacraments, for baptism, confirmation, extreme unction for the ordinations and religious dedications. In short, the Roman Catholic begins and ends life with an unction of the holy oil.

In the life of the ancients, a friction with perfumed oil was a hygienic practice followed quite generally. The athletes were rubbed with oil before appearing in the arena, so as to give more suppleness and vigor to their bodies, and this salutary usage began to be gradually abandoned only when the admiration for physical force ceased to enjoy favor among the people.

Bertile says that the elasticity and vigor that were found among the Grecians and Romans, were due, undoubtedly, to the use of olive oil, which was so popular among them. While animal fat is injurious to the stomach, and thins the blood, olive oil helps the digestion, enables the body to develop more suppleness, and helps the brain to attain the highest possible stage of human intellect. The salutary effects of olive oil on the human system have never been disputed.

The oil was also, and is yet, the basis of many perfumed preparations, and, as ladies of fashion and buxom dandies belong to all ages and to all countries, the use made of olive oil in that direction is not of an uncommon importance.

The fatty oils of low grades, either in their crude state or admixed with different preparations, are used also in considerable amount in soap-making, in lubricating, in lighting, in dyeing, in the manufacture of broadcloth, and they enter in the composition of many ointments and liniments.

It seems unnecessary to dwell on the great importance of the olive oil for table use. In the culinary point of view, it was of the very first necessity among the ancients, where oil cooking was predominant, and where it entered into all the seasonings most generally employed. This practice has happily been transmitted to us, and the use that is made of it nowadays in culinary preparations, sauces, salads, etc., is sufficiently demonstrated by its immense annual production, in which Italy alone figures for about 92,000,000 gallons.

In Spain, where olive oil is the principal seasoning in culinary preparations, enormous quantities are consumed. Italy and Portugal use also a great deal of it in their cooking.

But it is especially in the south of France that oil cooking predominates. The inhabitants of those regions have but little love for butter and entertain a very moderate esteem for the culinary art of northern people. Let him that has not traveled in those favored sections and tasted their delicious cooking throw me the first stone! I was born there; from my early infancy I was fed on that most excellent and nutritious kind of cooking; I have kept it up through most of my life and feel happy to transmit it to my children who like it as much as I do. How often we permit ourselves to enjoy an innocent and pleasant joke towards the guests who sit occasionally at our modest table. I order for instance an omelet cooked with oil in place of butter. I keep this from my guest; I watch his countenance; he tastes it; "by Jove!" exclaims he, "what a fine omelet!" and I reply with an insinuating smile: "Oil cooking my friend!"

It was to supply the place of good oil, whose production was beginning to fall behind the consumption that the use of butter was introduced and became more and more general as the adulterations of the oil became more and more frequent. It is thus that those sophistications gained many proselytes to the cause of butter; but let us produce a strictly pure olive oil in California, where we have to help us to it a most exceptional soil and climate, we will gain back many followers to the old cause, and, in view of the enormous demand we have to meet in the United States alone, which will keep increasing all the time, and for which we have a protective customs duty of 25 per cent ad valorem on the foreign article, many generations will pass before we will find it necessary to compete in other countries with the European oil, whose production of the pure article, as already said, is not up to the actual consumption of the whole world, and which fact accounts for its many adulterations with cotton, sesame, poppy, cocoanut, lard oil, etc., when it is not something worse.

61

CHAPTER VIII.

THE PICKLED OLIVE.

The pickled olives appear in Europe on all tables as *hors d'œuvres*, or side dish, as an aperitive condiment; and the culinary art knows how to employ them in a thousand different ways.

In the United States they are found in the French, Italian and Spanish restaurants with a few exceptions, as also on the tables of the wealthy classes who, having traveled abroad, have learned and adopted this most pleasant habit. They are also found quite extensively in the best bar-rooms, where they are offered to consumers with the traditional cracker so as to predispose them to enjoy the drink they are going to imbibe.

They are a great resource for the poorer classes of the old countries, and in the southern regions of Europe they are still one of the principal elements of their sober alimentation. A piece of bread under his arm, a flask of wine and a pocket full of olives, such is the equipment for the noon meal that many laborers carry away with them to the field where they are going to spend the whole day.

The pickling of the olive is a very simple operation. This is the method recommended by Coutance:

"The celebrated olives pickled after the manner of Picholini are submerged in a strong lye rendered more alkaline by an addition of quick lime. After leaving them in it for a certain time, which depends on their size, on the strength of the lye, and which is to be limited to the moment the pulp is penetrated to the pit, they are withdrawn, washed, and kept afterwards in water, to which is added about ten per cent of its weight of salt."

This is the mode given by Du Breuil: "Among the several receipts in use to take away the bitterness of the olive, we will indicate the one which we owe to the brothers Picholini of Saint Chamas, and which is considered the best: The olives are picked from the tree when they have reached their full development, but when they are still green, which is about the middle of September. They are dipped in a strong lye of potash, where they are left until the flesh is penetrated to the kernel. The lye is then replaced by fresh water which is removed twice a day during the first five days; after this they are kept in a strong brine."

In Bernays we find also the following recipe: "The method of preparing picholines in France, consists in putting the olives into a lye made of one part of quick lime to six parts of ashes of young wood sifted. After having left them half a day in this lye, they are taken out of it and put in fresh water, where they are allowed to remain eight days, the water being carefully renewed every twenty-four hours. After this a brine is made of a sufficient quantity of marine salt dissolved in water, to which is added some aromatic plants."

Here is now a process which is mostly the repetition of those I have just given, but which contains a few additional particulars which have come under my observation while pickling olives in Europe as well as here;

In the first place, the strength of the lye in which the olives are to be submerged has to be regulated. To that end I have employed the "American Concentrated Lye," which is found here at all groceries, in a solid state, in one pound boxes. After breaking the tin envelope I dissolve this cake of concentrated lye in a wooden bucket, into which I throw one gallon of hot water. When fully melted I have a lye of 13° to 14° strength, measured by the Beaume's hydrometer, which can be had at such hardware stores as Justinian Caire, of San Francisco, who imports them from Europe. With such a degree of strength the flesh of the olives is penetrated to the kernel in about five hours, which can be easily ascertained by taking one of them every five or ten minutes, after the first four hours, and cutting a slice from it with a pen-knife. The moment the flesh is fully penetrated I draw off the lye and I replace it by fresh water, which I renew in its turn five or six times at intervals of from six to eight hours. This renewal of water has for effect to clear the olives from the taste of the lye. Still, as they retain yet a little bitterness, it is finally removed by placing them for two or three days in a brine prepared on the basis of ten per cent of marine salt. Wild laurel leaves being thrown in this brine, will impart a delicious flavor to the olives, which are then ready for market. Whilst transferring them to bottles or barrels for shipment, these packages should be well filled with a new brine of the same strength.

There are a few other points in connection with this which I consider it important to follow.

1. Pick only from the tree the well developed berries that are perfectly green, and have not commenced yet to turn to a purple color. This can be done here in September, or the very latest, early in October. By waiting later they would be spotted by the oil forming in them, and would be unfit for the trade, though just as good for private consumption.

2. The pickling operations should be done only in wooden vessels, and rubber gloves should be used when the hands have to come in contact with the lye.

3. The lye should be left to settle as completely as possible before covering the olives with it, otherwise the strength of its sediment would spot many of them.

4. The olives should be covered with sacks or straw, with stones above, in order to keep the top ones from floating, in which case they would turn black.

5. The vessels should be so disposed as to allow the lye to be drawn off rapidly and completely, otherwise by too long a contact with this strong lye, some of the olives would be spotted or would turn soft.

While operating on large quantities, the wooden troughs should be disposed in such a manner that the same lye can be used in turn for all the olives that are to be pickled, provided, however, it is drawn every time in a separate trough where its strength can be regulated by a slight addition of fresh concentrated lye of a higher degree, and care taken that it settles well before using it again.

It can thus be seen that the pickling of the olive is a very simple, very rapid, and very cheap operation. The more so, as the moderate expense of making the lye, of which a small quantity covers great many pounds of olives, can be brought down nearly to nothing by its use, or its sale; as a winter tree wash, for it happens to be the very best preparation that can be used to that effect for ridding fruit trees of the numerous insects that live or deposit their eggs on them.

We can thus safely claim that nothing, or next to nothing is lost in the transformation of the product of the olive tree into a trade article.

CHAPTER IX.

CONCLUSION.

In preparing for the public this brief treatise on olive culture, written from a Californian point of view, it was my object to enable agriculturists and capitalists, who desire to avail themselves of the unique advantages it has over any other culture, to form a correct idea of its general features, from the choice of the land most suitable for the olive tree to the marketing of its product.

With this in view I thought it better to avoid lengthy demonstrations, or superfluous details, such as abound in some agricultural publications, the greater part of which is generally filled with diffuse and extraneous matter, which causes the reader to glance hurriedly from page to page, and to reach the last without having noticed what there can be of real interest in them.

I also found it necessary to consult the works of the best known writers on olive culture, and to quote them freely, placing them side by side with my personal observations, so as to add the weight of their acknowledged authority to my own statements. I thus hope that this treatise, which combines the best foreign and home experience, and which I have endeavored to make brief, clear and concise, will be instrumental in helping, to a certain extent, the development of olive culture in California, for it presents advantages that may be looked for in vain in any other agricultural pursuit.

Columelle knew what he was about when he proclaimed the olive tree "the first of all trees," and Parmentier felt himself well justified in saying many generations after, " of all trees that the industry of man has made profitable, the olive tree deserves, without contradiction, the very first place." I, therefore, consider it unnecessary to dwell any longer on a point on which all the best agriculturists, ancient and modern, fully concur, and I will confine myself to passing briefly in review the main reasons, given more extensively in the previous chapters, that contribute to give it this universal reputation.

In the first place the hill, or mountain lands, dry and rocky, which appear to be the most propitious for the robust constitution of the olive tree can be bought in California at prices ranging much below those necessary for the culture of other fruit trees or vines.

The cost of planting on such lands and care of the trees during the first year will hardly reach $5 per acre; the purchase of one year old rooted cuttings will not exceed from $10 to $15 per acre, and the annual care will be less than $5 per acre until the trees come to bearing, in four or five years after planting the rooted cutting.

The machinery and appliances for pickling the olive and for making the oil are of an extreme simplicity. Both operations can be done in a very short time and they are so easy that no farmer, with ordinary cleanliness and care, can fail in turning out as good a product as obtained anywhere else; while this is far from being the case in wine making which requires special knowledge, as well as long and tedious care before the product is in a satisfactory condition to be sold.

The gathering of the olive berries can be done gradually from November until March. By allowing them to dry in the barn, weeks can elapse before extracting the oil from them, which will enable a farmer to attend meantime to more pressing work; but, if he so prefers, he can do it at once. Moreover if he has no oil crusher and press, he can ship his olives in sacks or boxes to any distance at a moderate rate of transportation, considering the value of the product under a small volume, thus avoiding the misfortune of becoming the prey of local monopolies. How different it is with grapes! They are to be picked hastily when ripe; they must be pressed within a very short time; they cannot remain long, nor travel far without experiencing damage and loss; and if they are to be shipped to some distance to avoid the tyranny of monopolies, or because there is no wine cellar near by, the cost of freight, drayage, brokerage, short weight, added to the cost of picking and delivering absorb a good part of the value of a product which sold last year at an average of $20 per ton, and which is most likely to sell much cheaper this coming season.

On an equal acreage, and when from eight to ten years old, the product of an olive grove will be worth several times that of a vineyard; and under the same volume the oil will be ten times more valuable than wine, so that it can be delivered in a more economical manner. While with a four horse team a farmer will deliver about 600 gallons of wine per trip, representing a maximum value of $100, he can, with the same team, deliver olive oil to a value of over $1000. What an economy this represents!

Much less cooperage, too, will be required. Whereas, for a hundred acres vineyard, room for 50,000 gallons might be calculated upon, 25,000 gallons will be all that can be expected from a similar acreage of olive trees, and as tin tanks and cans are mostly used, it will cost less. Moreover, oil can be made from November to March, and sold shortly afterward to the merchant, who will clarify it himself, so that by spreading over the time of making it, a maximum of 8,000 or 10,000 gallons of such packages will be sufficient. And all this can be done and stored in wooden buildings of very moderate size, while a wine cellar should be built with stones or bricks, or be exposed to the danger of having the wine damaged or spoiled during the summer months, if it has not been sold before that time.

The gathering of the olive crop, too, is a very easy and cheap work. The berries that have fallen to the ground are first picked, then the tree is shaken and the branches struck with long poles to cause the fall of the remaining fruit. The few of them that may be found a little moulded, by a too long contact with the earth, though good enough to make good oil, are generally set apart to be used only with the last pressures, when the lower grade of oil is made. Let us compare this easy and rapid work, where nothing is lost, with the picking of grapes, or the product of most of fruit trees, which necessitates a certain number of hands at a given time, and requires special care, so as not to spoil part of it, while the fruit found on the ground is not marketable, if not entirely worthless.

When the oil is made, the residues, or marcs, are used for fuel, manuring, or feed for horses and cattle. There is, thus, not a farthing's worth of value in the product of the olive tree that is not turned to some use.

The bitterness of the fruit of the olive, of its bark and leaves, offers by itself a certain amount of protection against the attacks of insects and animals; and, when the tree is planted on hills, where it should be, far from moist places which engender most of the diseases of fruit trees, it has not to dread such terrible enemies as those that assail the vine, from the Oidium to the Phylloxera, which, alone, within the last twenty years, has brought down the French wine production from 85,000,000 hectolitres (about 2,000,000,000 gallons) to 25,000,000 (about 625,000,000 gallons) and which crops slowly and relentlessly on among our California vineyards.

During the excessively dry summers which are occasionally seen in part of California, when all the other agricultural productions are affected and diminished in consequence, the olive tree, this king of the dry soils, where it vegetates best, will continue to be loaded with fruit, just as in the seasons most favorable to other cultures.

The spring frosts, so disastrous generally to valley land vineyards, seem to have no effect on the olive. The tree is often affected and even killed in the best oil regions of Europe by excessive cold spells, which are absolutely unknown in our parts of California, so that its culture, which offers great danger there, and keeps it from being more developed, presents an unquestionable safety in Napa Valley and such other sections where there is no danger of such extremes of cold or hot weather, both of which the olive tree fears to an equal degree.

Finally, while an olive grove planted with one year old rooted cuttings pays, when five and six years old, quite as much as a vineyard of same age; twice as much when from seven to eight years old, and increases from year to year its annual paying power to $300, $400, $500, per acre, and upwards, until, when about twelve to fifteen years old, the tree reaches its full bearing capacity, on what basis shall we calculate then the cash value of such an orchard? Were I to mention between $1,500 and $2,000 per acre many people not fully acquainted with this culture would consider it a gross exaggeration. If such orchards are worth over $1,000 per acre in Europe, where olive trees are liable to be frozen at frequent intervals, why should they not be worth more here on account of the absolute immunity of those trees against such danger? Do not also protective duties insure us better prices for our oil as they do for our wines? Should import duties ever be abolished on both products, which would suffer most; the oil that pays only 25 per cent. on its value in the European markets, or the wine that pays 50 cents per gallon, which is more than double the value of the ordinary wines in France? We will thus see those prices of $1,500 and $2,000 per acre in California when the young olive orchards planted within the last few years shall have given the full measure of their worth. They will confirm by their development the careful demonstrations I have endeavored to make in this work.

By adding to what precedes the incredible longevity of the olive tree and the immense consumption that is enjoyed by its product in all the civilized parts of the world, it will be readily understood why Columelle, Parmentier, and so many other famous agriculturists of past and present generations have called it "The first of all trees," and why the Italians, whose oil production exceeds that of any other country, have popularized the proverb that we should never tire repeating in California; "*An olive plantation is a gold mine on the surface of the earth.*"

CONTENTS.

Chapter I—Soil................................... 5
" II—Reproduction..................... 11
" III—General Care..................... 25
" IV—Cost of a Plantation............. 31
" V—Diseases 35
" VI—Varieties........................... 41
" VII—The Olive Oil..................... 53
" VIII—The Pickled Olive............... 63
" IX—Conclusion......................... 69